DATE DUE JUN 1 1 2016

JAN 0 7 2003		
MAR 1 9 2007		
SEP 0 7 2007		
MAR 1 8 2009		
MAY 1 9 2009		
MAR 1 2 2010		
JUN 2 2 2011		
JUN 2 3 2011		
JUN 2 9 2013		

Demco, Inc. 38-293

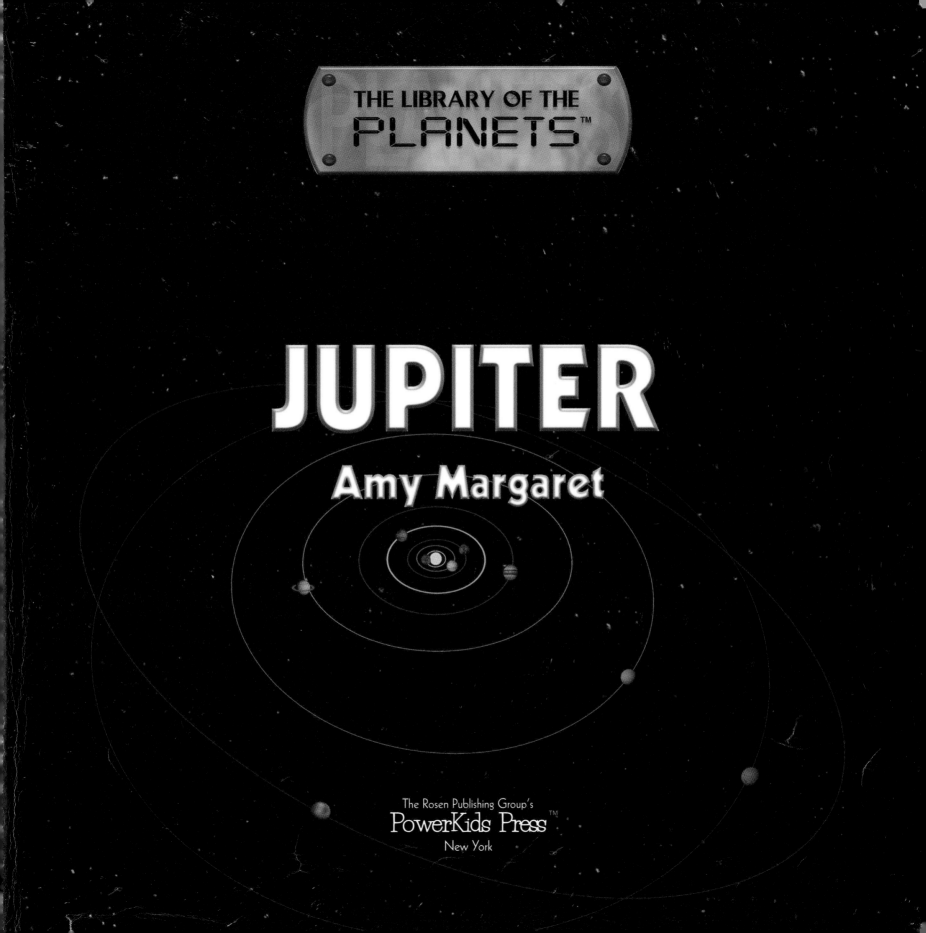

THE LIBRARY OF THE PLANETS™

JUPITER

Amy Margaret

The Rosen Publishing Group's
PowerKids Press™
New York

For Brittany, Prest-O, and Megan

Published in 2001 by The Rosen Publishing Group, Inc.
29 East 21st Street, New York, NY 10010

First Edition

Book Design: Michael Caroleo and Michael de Guzman

Photo Credits: pp. 1, 4 PhotoDisc; p. 7 (Roman god Jupiter) Michael R. Whalen/NGS Image Collection, p. 7 (Jupiter) PhotoDisc (digital illustration by Michael de Guzman); p. 8 © CORBIS; p. 10 PhotoDisc (digital illustration by Michael de Guzman); pp. 11, 12, 16 Courtesy of NASA/JPL/California Institute of Technology; p. 15 PhotoDisc (digital illustration by Michael de Guzman); p. 19 Courtesy of NASA/JPL/California Institute of Technology (digital illustration by Michael de Guzman); p. 20 © Dewey Vanderhoff/Liaison Agency.

Margaret, Amy.
    Jupiter/ by Amy Margaret.
        p. cm.— (The library of the planets)
    Includes index.
    Summary: Describes the history, unique features, and exploration of Jupiter, the
fifth planet from the Sun.
    ISBN 0-8239-5647-4
    1. Jupiter (Planet)—Juvenile literature. [1. Jupiter (Planet)] I. Title. II. Series.

QB384 .M35 2000
523.45–dc21   99-089526

Manufactured in the United States of America

Contents

Jupiter, the Giant Planet

Jupiter is the largest of the nine planets in our **solar system**. A solar system is a group of planets that move around a star. In our solar system, that star is the Sun. Jupiter is the fifth planet from the Sun.

Jupiter is one of the **gas giants**. The other gas giants are Saturn, Uranus, and Neptune. The gas giants are made mostly of liquid and gas. Most of Jupiter is made from the liquid form of the gas hydrogen. Scientists believe that Jupiter has a rocky **core** made of iron.

Jupiter and the other planets in our solar system were most likely formed about 4.5 billion years ago. Many scientists believe that our whole **universe** was formed during the Big Bang. The Big Bang was a huge explosion that took place about 10 billion years ago. About 45 million years ago, the planets in our solar system began to form out of a hot cloud of gas and dust.

Jupiter is the largest of the nine planets in our solar system. It is made mostly of hydrogen, which is a gas. Jupiter's inner core is believed to contain iron.

The History of Jupiter

The planet Jupiter was named after the king of the gods in Roman **mythology**. Jupiter is certainly "king" of our solar system in size. It is twice as big as all the other planets in our solar system put together. It is so large that if it were hollow it could hold 1,300 planet Earths! Even three Saturns could fit inside one Jupiter, and Saturn is the second largest planet.

In 1610, Jupiter was seen through a **telescope** for the first time. A telescope is an instrument used to see objects that are far away. The **astronomer** who saw Jupiter was named Galileo Galilei. Galileo also discovered Jupiter's four largest moons. They were named the Galilean moons after Galileo. Jupiter is very far away from Earth. It was hard to study Jupiter until recently, when scientists began using **space probes**. A space probe takes pictures of objects in space and sends them back to Earth through computers.

In Roman mythology, Jupiter, the king of the gods, is often shown holding lightning bolts in his hand. This is why the symbol for the planet Jupiter is a lightning bolt.

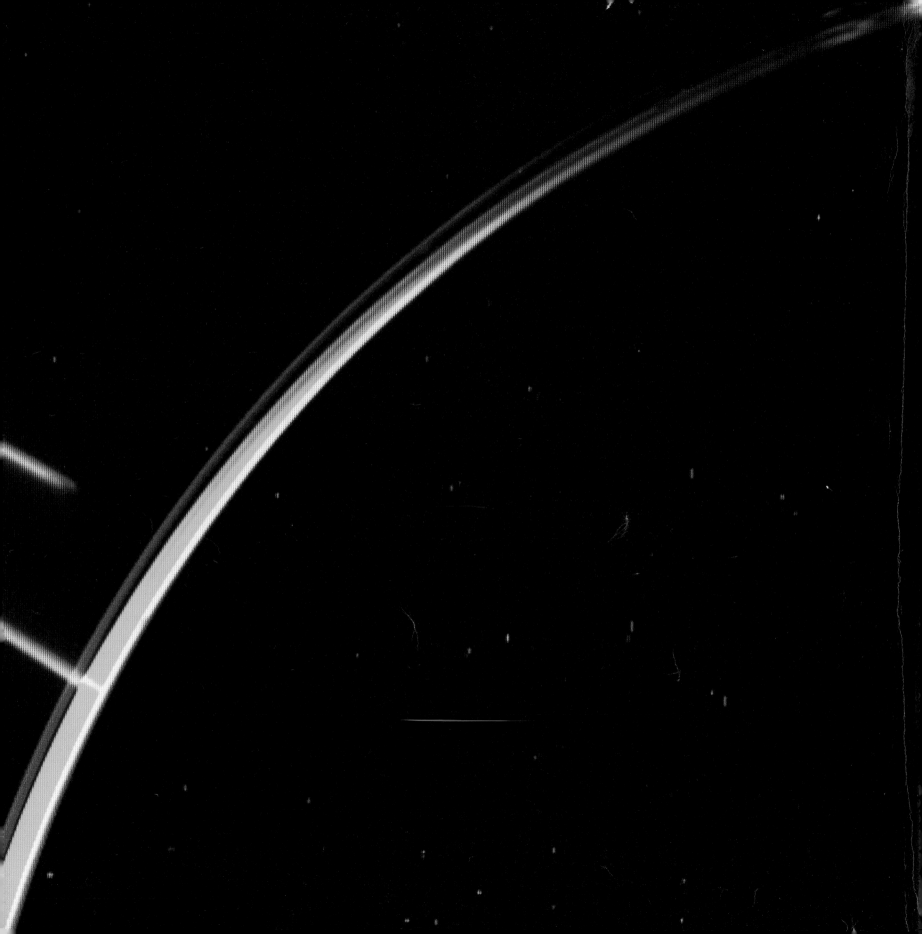

Jupiter's First Visitors

In 1972, the space probe *Pioneer 10* was launched into space. A space probe does not carry passengers. It is steered by scientists on the ground. *Pioneer 10* passed Jupiter on December 3, 1973. *Pioneer 11* was launched in 1973. The two *Pioneer* probes sent back information about the temperature of Jupiter's **atmosphere**. Both probes also sent back pictures of Jupiter's moons.

The next mission was a pair of space probes called *Voyager 1* and *Voyager 2*. These probes passed Jupiter in 1979. They made many new discoveries. They proved that lightning existed on Jupiter. They also photographed three new moons. One of the biggest surprises was the discovery of rings around Jupiter. During the *Voyager 1* and 2 missions, pictures of the planets Saturn, Uranus, and Neptune were taken. *Voyager 1* and 2 will send back information about outer space until the year 2015.

The space probe Voyager 2 took this picture of Jupiter's rings in 1979.

The Parts of Jupiter

Jupiter does not have a solid **surface** like Mercury, Venus, Earth, and Mars. Its atmosphere is made mostly of the gas hydrogen, along with a few other gases. The atmosphere is the layer of gases that surrounds a planet.

In between Jupiter's outer atmosphere and its rocky core is a thick layer of liquid hydrogen. From our view on Earth, Jupiter's surface looks like it is dotted with light and dark ovals. These ovals are cloud systems. The cloud systems move in circles. The dark areas mean that

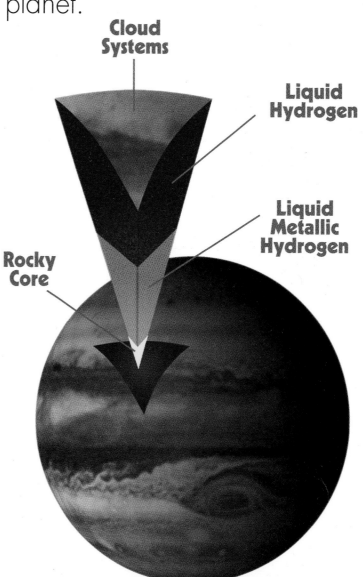

Cloud Systems

Liquid Hydrogen

Liquid Metallic Hydrogen

Rocky Core

there are smaller amounts of a gas called ammonia in those parts of the cloud system.

A powerful telescope called the Hubble Space Telescope took these pictures of the comet Shoemaker-Levy 9 as it hit Jupiter in 1994. The image at the bottom right was taken more than one month after the one at the top left.

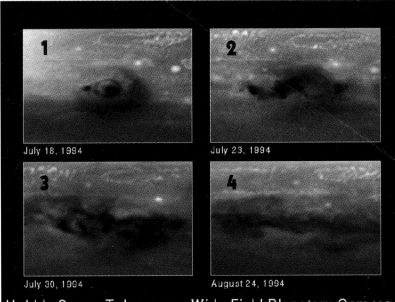

1 — July 18, 1994
2 — July 23, 1994
3 — July 30, 1994
4 — August 24, 1994

Hubble Space Telescope • Wide Field Planetary Camera 2

The Comet Crash

A comet is an icy ball that orbits the Sun. Several pieces of a comet struck the planet Jupiter in 1994. If you see photos of Jupiter before the comet hit and photos taken after it hit, you can see the effect the comet had on Jupiter. The comet was named Comet Shoemaker-Levy 9, after the two people who discovered it.

An icy layer of clouds covers most of the hot liquid hydrogen that makes up Jupiter. The temperature of the rocky core is about 36,000 degrees Fahrenheit (20,000 degrees C).

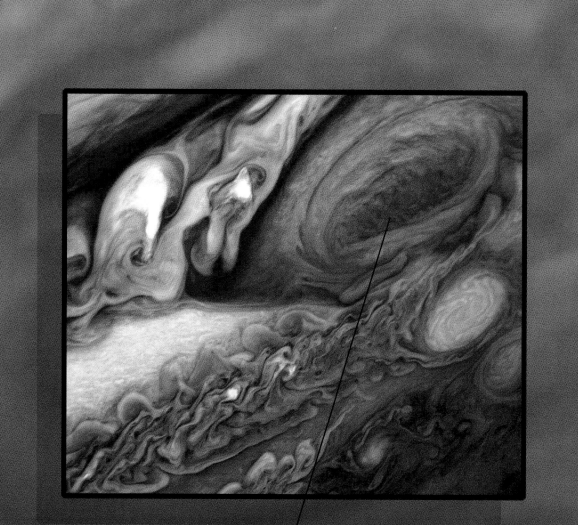

The Great
Red Spot

The Great Red Spot

One of the most interesting things about Jupiter is its Great Red Spot. The Great Red Spot is a huge storm that has been moving across Jupiter's surface for at least 300 years.

The Great Red Spot, which is up to 25,000 miles (40,225 km) long, may look small on the planet Jupiter, but it is actually almost as wide as Earth.

Scientists are not sure why this storm has continued for so many years. The Great Red Spot needs energy to keep moving. It is possible that the Great Red Spot can get energy from smaller storms.

Voyager 1 took this image of the Great Red Spot of Jupiter. This photo is actually three different pictures that were taken by special cameras and then put together by scientists.

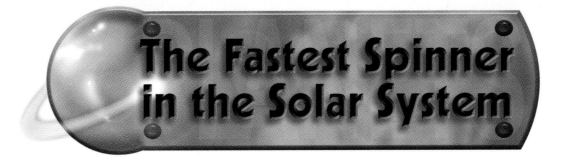

Jupiter and the other eight planets are very different from one another. What they all have in common is the way they move. First, each planet spins on its **axis**, like a merry-go-round. Jupiter spins on its axis quicker than any other planet. It makes one complete **rotation** in less than 10 hours. It takes Earth 24 hours, or one day, to rotate once on its axis.

The nine planets also circle the Sun. The force of the Sun pulls the planets close to it. The closer a planet is to the Sun, the faster it travels around the Sun. Jupiter moves much more slowly than Earth. Earth circles the Sun in 365 days, or one year. It takes Jupiter 12 Earth years to travel around the Sun! This is because Jupiter is much farther from the Sun.

Jupiter makes one rotation on its axis in 9 hours and 51 minutes. It takes Jupiter about 12 Earth years to make one complete rotation around the Sun.

THE SUN

JUPITER

Planet	Orbit Time Around Sun
Mercury	88 Earth Days
Venus	225 Earth Days
Earth	365 Days
Mars	687 Earth Days
Jupiter	12 Earth Years
Saturn	29 Earth Years
Uranus	84 Earth Years
Neptune	165 Earth Years
Pluto	249 Earth Years

FUN FACTS

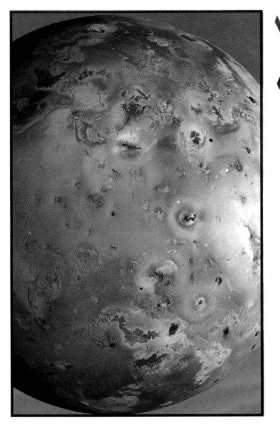

The Galileo spacecraft took this picture of Jupiter's volcanic moon Io. Io is about the same size as Earth's moon.

Unlike Io, the moon Europa has a smooth surface. Ganymede is the largest moon in our solar system. Callisto has a crust of ice, just like Ganymede.

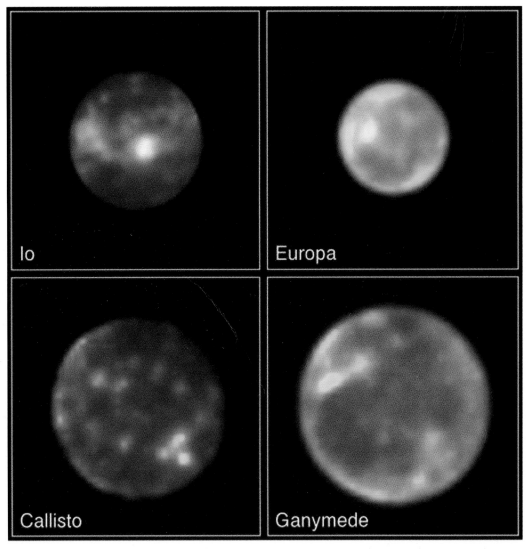

Io

Europa

Callisto

Ganymede

Jupiter's Many Moons

Jupiter has 16 moons. The four biggest moons are Io, Callisto, Europa, and Ganymede. Ganymede is the largest moon in the entire solar system. Callisto is a **crater**-covered moon. Most of the craters are larger than a half mile (1 km) in size. Some scientists think that a saltwater ocean might lie underneath the surface of Callisto. Io contains active volcanoes. Io is the first place besides Earth where scientists have found active volcanoes. Some of its volcanoes are as hot as 3,140 degrees Fahrenheit (1,727 degrees C). Even Venus, the hottest planet in the solar system, gets only as hot as 900 degrees Fahrenheit (482 degrees C). Scientists think that life may exist on the fourth moon, Europa. Only three things are needed for life to survive. These things are heat, liquid water, and living material. All three of these things have been found on Europa. Europa will be the main subject of the next U.S. space mission to Jupiter.

Jupiter's Rings

Jupiter has a thin set of rings. They were discovered in 1979 by *Voyager 1*. The rings seem to be about a half mile (1 km) thick. Scientists are not sure what elements make up these rings. They could be pieces of ash from the moon Io's volcanoes. They could also be made of dust from objects in space, such as a comet. The rings circle Jupiter about 30,000 miles (48 km) above the top of the planet's atmosphere. The rings of Jupiter are dark. They are much darker than Saturn's bright rings.

A special camera on the Hubble Telescope took this view of Jupiter's moon Metis, one of Jupiter's rings, and the clouds above the planet.

The Ringed Planets

Only four of the nine planets in our solar system have rings. They are Jupiter, Saturn, Uranus, and Neptune.

If you weigh 100 lbs. (45.4 kg) on Earth, you would weigh 254 lbs. (115 kg) on Jupiter.

FUN FACTS

To find out where Jupiter is in the night sky, look on the Internet. These addresses will help you figure out where it is:

http://www.kidsnspace.org/what_can_i_see.htm
http://www.skypub.com/sights/sights.shtml

You can see what the *Galileo* spacecraft is doing every day on this Web site:

http://www.jpl.nasa.gov/galileo

Mercury

Jupiter

Venus

Seeing Jupiter in the Night Sky

Jupiter is one of the few planets you can see from Earth. You do not even need a telescope. Jupiter appears at night. It appears in different places in the sky at different times of the year. Check a monthly **astronomy** magazine or the Web sites listed on page 20 to find the best time to see Jupiter.

If you have a telescope, you may be able to see Jupiter's four largest moons. The moons of Jupiter move around the planet at different speeds. This means that every night when you look at Jupiter, you will see the moons in different spots.

Jupiter is the fifth planet from the Sun. Mercury is the closest planet to the Sun, Venus is the second closest, and Earth is the third.

The Great Galileo Mission

The *Galileo* spacecraft was sent into space in 1989. It arrived in Jupiter's atmosphere in 1995. One of its most important jobs was to release a small probe near Jupiter. Scientists who built the probe knew that it would survive only a short time because the atmosphere near Jupiter was so hot.

The probe measured the winds on Jupiter and did other tests. It sent back the information through computers. Within an hour, the hot temperatures on the planet destroyed the probe.

Galileo is still orbiting Jupiter. It has taken pictures of more than 100 volcanoes on Io. *Galileo* has also taken pictures of the Great Red Spot, and other moons that orbit Jupiter.

There is still much that we don't know about Jupiter. Future spacecraft research missions to Jupiter may help us discover more about the largest planet in our solar system.

Glossary

astronomer (uh-STRA-nuh-mer) One who studies the night sky and the planets, moons, stars, and other objects found in space.

astronomy (uh-STRA-nuh-mee) The science of the sun, moon, planets, and stars.

atmosphere (AT-muh-sfeer) The layer of gases that surrounds an object in space. On Earth, this layer is air.

axis (AK-sis) A straight line on which an object turns or seems to turn.

core (KOR) The center layer of a planet.

crater (KRAY-ter) A hole in the ground shaped like a bowl.

gas giants (GAS JY-antz) Planets made up mostly of gas.

mythology (mih-THA-lih-jee) The stories that people make up to explain events in nature or people's history.

rotation (roh-TAY-shun) The spinning motion of a planet around its axis.

solar system (SOH-ler SIS-tem) A group of planets that circle a star. Our solar system has nine planets, which circle the Sun.

space probes (SPAYS PROHBZ) Small spacecraft that travel in space and are steered by scientists on the ground.

surface (SER-fis) The top or outside of something.

telescope (TEL-uh-skohp) An instrument used to make distant objects appear closer and larger.

universe (YOO-nih-vers) Everything that is around us.

Index

Web Sites

If you would like to learn more about Jupiter and the other eight planets, check out this Web site:
http://seds.lpl.arizona.edu/nineplanets/nineplanets/nineplanets.html

24